Familiar Butterflies

Alfred A. Knopf, New York

Prepared and produced by Chanticleer Press, Inc., New York.
Typeset by Dix Type, Inc., Syracuse, New York.,
Printed and bound by Toppan Printing Co., Ltd., Hong Kong.

Published June 1990
Sixth printing, October 1997

Library of Congress Catalog Card Number: 90-052502
ISBN: 0-679-72981-X

Contents

Introduction

The Butterflies

Appendices

How to Use This Guide

Butterflies occur in virtually all areas of North America. Learning to identify common butterflies will lead you into a world of dazzling colors and fascinating behaviors.

Coverage

This new guide describes and illustrates 80 of the most abundant and widespread butterflies in North America. Additional similar or related species are also mentioned, broadening the scope of the book.

Organization

This easy-to-use pocket guide is divided into three parts: introductory essays; illustrated accounts of the butterflies; and appendices.

Introduction

As a basic introduction, the essay "Identifying Butterflies" tells you what characteristics to look for when you see an unfamiliar butterfly. "Watching Butterflies" gives you practical clues for observation.

The Butterflies

This section includes 80 full-page color illustrations of the butterflies and skippers covered in the guide. They are arranged in six main groups that are presented in the same order in which they are classified by lepidopterists. On the page facing the photograph is the text account. It begins with an introductory paragraph that provides facts about each species' biology, habits, or history. Next comes a description of the important field marks of the species, as well as information about its life cycle, habitat, and range. Many butterflies have significantly different appearances when their wings are spread and when they are perching with them closed; therefore we have included, in most accounts, a small photograph of the closed (ventral) views on the text page. There are a few butterflies that only perch in one position; here we have included instead a small photograph of a similar species, a seasonal form, or a variation in color for purposes of comparison. Usually the description covers a single species; at times, similar or closely related species are also mentioned.

Appendices   Many related species are broadly similar in appearance or behavior. To help you recognize a butterfly as a member of a particular group, we have included a special section, "Butterfly Families." Here are described the major groups of butterflies included in this guide. Black-and-white drawings show the basic parts of butterfly anatomy. Also included are a glossary, defining commonly used terms, and an index.

7

Life of a Butterfly

Butterflies go through a complicated life cycle known as metamorphosis. After mating, female butterflies lay eggs either singly, or in rows, chains, or clusters of a few to several hundred eggs. The egg shape and texture varies greatly among species, ranging from spherical, flattened, or conical to smooth, ribbed, or ornamented with raised designs. Some eggs do not hatch until the following spring, while others hatch before winter. The caterpillar, or larva, has simple eyes, chewing jaws, and three pairs of jointed legs near the front as well as five pairs of grasping prolegs near the rear. The caterpillar spends its life feeding; the more it consumes, the larger it grows. But because its skin cannot stretch, the caterpillar grows by molting or shedding its skin several times—each stage, called an instar, is larger than the previous one. The final molt produces the chrysalis, or pupa; during this stage it does not feed. The chrysalises of most butterflies are naked, unlike those of moths, which are protected by a silken cocoon. Butterfly caterpillars can produce silk, which they use to bind leaves together for a shelter, which may be used either by the caterpillar or later by the chrysalis. The chrysalises of many butterflies hang by the tail end from a silken pad called the cremaster; others hang upright, supported by a silken girdle. Some chrysalises are green or brown, resembling leaves, stems, thorns, or bits of wood. Others are brightly variegated and covered with thornlike bumps or tubercles. As the adult butterfly begins to form inside the chrysalis, the shape of the compacted wings as well as the features of the head and body are visible in the surface of the chrysalis. When the adult is fully formed, the chrysalis' skin splits out. It soon begins to pump fluids from its swollen body into its shrunken wings. This transformation from egg to adult butterfly is known as complete metamorphosis.

As soon as the adult can fly, courtship begins, sometimes involving elaborate dances, prenuptial flights, or mutual wing strokings. Mating usually lasts several hours, and often occurs while the pair is flying. Nearly every butterfly's lifespan ranges from a week to six or eight months, depending upon the species, with most averaging two weeks. In warm regions, several broods of a species are produced each year, but in mountains or the Arctic, it may require two years for a single brood to mature.

Identifying Butterflies

An important step in identifying an unknown butterfly is estimating its overall size. Is it relatively large, like a swallowtail or Monarch? Medium-sized, like the Painted Lady? Or small, like many skippers or blues? Establishing the small—medium—large categories relative to butterflies you know will help you to identify unfamiliar species.

Next, notice the most obvious overall colors and patterns. Species accounts often mention these characteristics first, and in combination with size they are often enough to make an identification.

Special features—such as tails, eyespots, stigmata, and wing shape—can also be useful in identification. Check the black-and-white drawings as well as the glossary to become familiar with these characteristics and terms.

You can frequently make an identification by eliminating one or another similar species. Many of the accounts in this guide mention such look-alikes, providing a brief account of how they differ from the featured species.

With experience, the observer will come to recognize the major groups of butterflies. Recognizing that an unknown species is a skipper, a swallowtail, a tortoiseshell, or an anglewing will simplify identification. Information in the section titled "Guide to Butterfly Groups" will help you get a start on this useful technique.

A Note on Scientific Names

The common, or English, names of butterflies are often colorful and evocative: Great Spangled Fritillary, American Painted Lady, Comma. Common names, however, are not standardized and show much variation. But every butterfly is assigned a distinct and unique scientific, or Latin, name, that is standard around the world and governed by an international set of rules.

A scientific name has two parts. The first tells us to which genus (plural, genera) a butterfly belongs; the second tells us the species. (A species is a kind of plant or animal that is capable of reproducing with members of its kind but is genetically isolated from others.) Most genera include many species. Sometimes a species includes two or more distinct groups (usually called subspecies) that differ in range, color, size, or some other factor. Subspecies are indicated by the use of a three-part scientific name. The White Admiral *(Limenitis arthemis arthemis)* and the Red-spotted Purple *(L. arthemis astyanax)* are subspecies.

The names of species and genera are frequently reviewed by lepidopterists. When species have formerly been identified by other scientific names, we have mentioned them in the text.

Watching Butterflies

One of the appeals of watching butterflies is the opportunity it affords for enjoying beautiful weather. Butterflies are, to a large degree, creatures of sunlight; calm, sunny days provide ideal conditions for observation. In the southern states, some species fly all year round. Northward they are creatures of the warmer months (March to October). You may want to start in a field or prairie with a good assortment of wildflowers, with at least some in bloom. Walk slowly along the edges or paths and also visit any obvious concentration of flowering plants. The chances are good that you will soon encounter one or more kinds of butterflies.

Approach perching or nectaring butterflies in a slow and steady manner. Dashing up to your quarry will, more often than not, end with the butterfly making a quick exit—and trying to outsprint these creatures is normally fruitless. With experience, you will improve your "search image," a facility for knowing how and where to look.

By visiting different habitats and plant communities, you will see a greater variety of species. Wetland edges, woodland glades, bogs, dune communities, chaparral, and tundra all have their associated butterflies. Take advantage of the diversity of habitats and plant communities in your area; and each time you go out to look for butterflies, try to visit one unfamiliar place. Land that has been altered by human activity may also be productive; powerline cuts, railroad rights-of-way, abandoned lots, and weedy wasteplaces are all areas where new growth attracts butterflies.

Another important factor to keep in mind when looking for butterflies is timing. Different species have different flight periods—times of the year when they are active. Walks in spring, midsummer, and fall will often yield different species.

One excellent way to record your observations is with a camera. A telephoto lens and a tripod will help you take revealing "candids" of the species you encounter. If you want a closer look, try carrying a net and clear plastic jar into the field. You can capture a butterfly, transfer it to the jar for observation, and then release it once you have had a good look. But if you plan to handle or capture butterflies, do so with great care. They are fragile beauties.

The Butterflies

Pipevine Swallowtail *Battus philenor*

Like the Monarch, the Pipevine Swallowtail serves as a model that other species mimic. Predators avoid not only the distasteful pipevine but other, presumably palatable, look-alikes; these include female Black Swallowtails, female Spicebush Swallowtails, and Red-spotted Purples. Similar flight behaviors as well as closely matched physical features have evolved in the mimics.

Identification	Wingspan 2¾–3⅜". Overall black. Above: iridescent blue on HW and border of FW; light dots along portions of the borders of the wings. Below: 1 row of large, orange spots in semicircle parallel to HW border.
Habitat	Open forests and woodland edges; also fields, meadows, roadsides, and orchards.
Range	S. New England south to Florida and Texas and west to California. Less common or absent in western plains states and the Northwest.
Life Cycle	Caterpillar feeds on pipevines (*Aristolochia* spp.). Adults, 2–3 flights per year.

18

Zebra Swallowtail *Eurytides marcellus*

The Zebra Swallowtail belongs to a large genus of tropical butterflies called Kite-tailed Swallowtails. Two seasonal forms of the Zebra Swallowtail are readily recognizable. Butterflies of the early spring broods (marcellus) are overall smaller and lighter in coloration than individuals of the later, summer flight (lecontei). Butterflies of the lecontei form have the typical, long, kitelike tails that give this group its name.

Identification	Wingspan 2⅜–3½″. Overall greenish white with black stripes. Above: HW with red and blue markings near very long tails. Below: HW with red band through middle.
Habitat	Woodland water courses, swamps, and marshes.
Range	S. New England west through southern portions of Great Lakes and south through E. North America to Gulf Coast.
Life Cycle	Caterpillar host plant, pawpaws (*Asimina* spp.). Adults 2–4 flights per year.

20

Black Swallowtail *Papilio polyxenes*

Formerly known as the Eastern Black Swallowtail, this species is similar to Baird's Swallowtail *(P. bairdii)*, sometimes called the Western Black Swallowtail. In the Black Swallowtail, the black pupil on the hindwing eyespot is centered and round; in Baird's, it is off-center and oblate. The caterpillars are unpalatable to predators because they ingest toxic oils from their host plants.

Identification   Wingspan 2⅝–3½". Overall black. Above: iridescent blue on HW (more pronounced in female); row of yellow markings on outer margin of wings; orange eyespot with centered, black pupil on inside HW edge. Below: 2 rows of orange to red markings on HW. A subspecies, *P. p. coloro*, has wide yellow band below.

Habitat   Open habitats including meadows, weed fields, gardens, parks, hillsides, and ridges.

Range   Rocky Mountains east from S. Canada to Gulf of Mexico.

Life Cycle   Caterpillar feeds on plants of the carrot family (Apiaciae). Adults 2–3 flights per year.

Anise Swallowtail *Papilio zelicaon*

This wide-ranging species is found from coastal areas to timberline. The adults are strong fliers and often elusive, and the Anise Swallowtail is known among butterfly enthusiasts as a difficult species to approach. The caterpillar, like other swallowtail larvae, has a pair of "horns" (osmateria) behind the head. When extended, the horns give off a rank odor that may deter predators.

Identification   Wingspan 2⅝–3″. Above: black with wide yellow bands on both FW and HW; HW with some iridescent blue and paired orange eyespots with dark centers—one on inside edge of each wing. Abdomen with 2 yellow stripes.

Habitat   A wide variety of habitats including waste places, roadsides, and vacant lots.

Range   Southern portions of W. Canada to Baja California. Rocky Mountains to the Pacific.

Life Cycle   Caterpillars feed on members of the carrot family (Apiaceae) and citrus plants. Adults, 1 to several flights.

24

Giant Swallowtail *Papilio cresphontes*

The three largest North American butterflies are the King Swallowtail *(P. thaos)*, the Tiger Swallowtail *(P. glaucus)*, and the Giant Swallowtail. The caterpillar occasionally causes damage to southern citrus orchards and is called the "orange dog" by citrus growers. Sometimes called *Heraclides cresphontes*.

Identification Wingspan 3⅜–5½". Large. Above: black with 2 rows of large yellow spots converging at the wing tips, creating a bow-shaped pattern; HW with yellow-centered, rounded tails. Below: largely yellow. Males have notched tip to abdomen.

Habitat Semi-open to open areas, including meadows, river courses, hillsides, and citrus orchards.

Range S. New England to Florida, west to Rockies; parts of Southwest and S. California. More common southward.

Life Cycle Host plants include various citrus (Rutaceae), including rue *(Ruta graveolens)* and common prickly-ash *(Zanthoxylum americanum)*. Adults, 2 flights; year-round southward.

Tiger Swallowtail *Papilio glaucus*

In 1587, John White, a colonist in the service of Sir Walter Raleigh, drew an illustration of the Tiger Swallowtail. His model was a butterfly he collected on Roanoke Island—then part of the colony of Virginia. Sometimes classified in the genus *Pterourus*.

Identification   Wingspan 3⅛–5½″. Overall yellow with black stripes and tails. Above: shows black border; HW with some bluish iridescence and orange eyespots. Below: row of separate yellow markings on outside edge of FW; most markings on HW border are orange (see *P. rutulus*). Dark female form occurs in some areas.

Habitat   Woodland edges, meadows, parks, and suburbs.

Range   Alaska and Canada southeast to Florida and Gulf of Mexico; east of Rocky Mountains.

Life Cycle   Caterpillars feed on a variety of deciduous trees, including willows (*Salix* spp.), cottonwoods (*Populus* spp.), wild cherries (*Prunus* spp.), and yellow-poplar (*Liriodendron tulipifera)*. Adults, 1–3 flights per year.

Western Tiger Swallowtail *Papilio rutulus*

Most butterfly taxonomists consider the Western Tiger Swallowtail and the Tiger Swallowtail to be separate species. Maps showing the geographical ranges of the two species show a fairly clear line of demarcation, although some individuals showing intermediate characteristics (known as "intergrades") exist in British Columbia. Overall, however, one species replaces the other (or fills a similar niche) in different areas.

Identification	Wingspan 2¾–3⅞". Overall yellow with black stripes and tails. Above: black border; HW with some bluish iridescence and orange eyespots. Below: row of yellow markings on FW outside edge forms a solid band; most markings on HW border are yellow.
Habitat	Suburban plantings, woodlands, canyons, and roadsides; near streams and other water courses.
Range	British Columbia east to Rocky Mountains and south to Baja California.
Life Cycle	Host plants are deciduous trees including poplars (*Populus* spp.) and willows (*Salix* spp.). Adults, 1–3 flights.

30

Spicebush Swallowtail *Papilio troilus*

The Spicebush Swallowtail is another butterfly that mimics the Pipevine Swallowtail. While the Spicebush is often found in woodland habitats, adults also frequent meadows and gardens, where they sip nectar from a variety of flowers. Nectaring provides many butterflies with energy during their short but busy reproductive life.

Identification
Wingspan 3½–4½". Overall black. Above: single row of white to blue spots on outer edges of wings; 2 pairs of orange eyespots on HW, 1 on inside trailing edge; bluish (female) or greenish (male) iridescence on HW. Below: with 2 curved rows of orange-red spots on HW.

Habitat
Open woods, woodland edges, fields, and meadows.

Range
E. United States from Canadian border through Florida and E. Texas; largely absent on the plains and prairies except where wooded.

Life Cycle
Larvae feed mainly on spicebush *(Lindera benzoin)* and sassafras *(Sassafras albidum)*. Adults, 2 flights northward, 3–4 to the south.

Checkered White *Pontia protodice*

In the late 18th century, John Abbot, an Englishman, came to the New World to gather insects for the collections of wealthy British patrons. Between 1790 and 1810, Abbot lived and worked in Georgia, where he made the first systematic study of North American butterflies. His early efforts were published in London in *The Natural History of the rarer Lepidopterous Insects of Georgia* in 1797. The first formal description of the Checkered White is based on an Abbot drawing.

Identification — Wingspan 1¼–1¾″. Overall white, often checkered with blackish patches. Females, and both sexes of early (spring) broods, have more dark patches.

Habitat — Weedy areas, fallow fields, abandoned lots, and other disturbed habitats.

Range — Canadian–United States border south; Florida west to Baja California. Absent or scarce in Northeast and Northwest; more common in Southwest.

Life Cycle — Caterpillars feed on mustards (Crucifarae). Adults have many broods—all year in some areas.

Veined White *Pieris napi*

Seasonal variations in color are fairly common among butterflies. This is generally true of the pierids, or "Whites," and in particular of the Veined White. Spring broods develop in the chrysalis during shorter days; they tend to be strongly marked below with the dusky venation (as seen at left), from which this species gets its standard English name. Later, summer broods may be pure white or mustard-colored below without dark veins. This species is sometimes called *Artogeia napi*.

Identification	Wingspan 1½–1⅝″. Above: mostly white. Below: light mustard to white, often with dusky scaling along veins.
Habitat	Cool, moist woodlands as well as forest openings and edges.
Range	Alaska across Canada to N. New York and New England, and around Great Lakes. Northwest and Rocky Mountains south to central California and Arizona in proper habitat.
Life Cycle	Caterpillars feed on mustards (Cruciferae). Adults, 1–2 flights.

36

Cabbage White *Pieris rapae*

The Cabbage White, a Eurasian species, was first observed in North America at Quebec City in 1860. S. H. Scudder reports that the butterfly had spread south as far as Florida and west to Colorado by 1886. This species is now well established throughout the United States and Canada. Like many "imports," both plant and animal, the Cabbage White has thrived in its new home—some say at the expense of native species.

Identification
Wingspan 1¼–1⅞". Above: white with charcoal wing tips; 1 FW spot in males, 2 in females. Below: HW and FW tip mustard.

Habitat
Wide ranging; common in agricultural areas, weed fields, meadows, and suburban gardens.

Range
Most of North America from Canadian tree line south.

Life Cycle
Caterpillars feed on mustards (Cruciferae). Adults, 2–8 broods.

Sara Orange-tip *Anthocharis sara*

The western Sara Orange-tip is a wide-ranging species with several distinct subspecies; these geographic forms may, in fact, be different species. A similar eastern butterfly, the Falcate Orange-tip *(A. midea)*, is distinguished by its hooked wing tips. The Falcate Orange-tip is somewhat unusual; it flies even when the sun isn't shining.

Identification	Wingspan 1¼–1¾". Above white with red-orange wing tips bordered in black. Below: HW has mosslike appearance.
Habitat	Coastal plains to forested mountaintops; desert canyons and arid ridges.
Range	S. Alaska to N. Baja California; east to Rockies. Falcate Orange-tip found from Wisconsin east to Massachusetts, south to Texas and Georgia.
Life Cycle	Caterpillars feed on mustards (Cruciferae). Adults, mainly 1 spring flight.

Clouded Sulfur *Colias philodice*

This species and the related Orange Sulfur (*C. eurytheme*) occur throughout much of North America. Most individuals are yellowish to orange overall, but a certain percentage of females are white. These so-called "alba" form females develop faster and have larger eggs, which seems to give them a reproductive advantage, especially in more northerly climates where flight periods are restricted. This species is also called the Common Sulfur.

Identification	Wingspan 1½–2″. Overall lemon-yellow. Above: black borders on the wings—in males the black border is solid, in females the border is interspersed with light markings. Below: 1–2 spots (one silvered) ringed in red and a row of dark spots along wing border.
Habitat	Open fields and meadows, especially with alfalfa or clover (Leguminosae).
Range	Throughout most of North America; absent from high Arctic, S. Florida, S. Texas, and arid Southwest.
Life Cycle	Caterpillars feed on legumes. Adults, 1 to many flights.

42

Orange Sulfur *Colias eurytheme*

The Orange Sulfur and the Clouded Sulfur *(C. philodice)* are closely related. In areas where both are populous, there is a significant amount of hybridization. The resulting offspring are partly orange and partly yellow on the upper wing surfaces. Some authorities suggest that individual butterflies showing any orange at all should be called Orange Sulfur *(C. eurytheme)*.

Identification Wingspan 1⅝–2⅜". Overall yellow. Above: bright orange with black wing borders—solid in males, broken in females. Below: HW with 1 or 2 red-ringed spots (1 silvery) and a row of dark spots along wing border. Alba females have pink wing fringe.

Habitat Open fields and meadows, especially with white sweet clover or alfalfa (Leguminosae).

Range Most of North America south of Canadian taiga and tundra.

Life Cycle Larvae feed on alfalfa and clovers (Leguminosae). Adults, 1 to many flights; from February–December in warmer climates.

44

Southern Dogface *Colias cesonia*

The two North American dogface butterflies are members of a largely tropical group of sulfurs. The Southern Dogface and its close relative, the California Dogface *(C. eurydice)*, both have curved and pointed tips to the forewings. The Southern Dogface can be distinguished by the dark markings on the upper wing borders. Also, the male California Dogface has an iridescent purplish sheen on the forewing. Both are sometimes included in the genus *Zerene*.

Identification	Wingspan 1⅞–2½″. Overall yellow and black. Above: yellow "dogface" design, like a poodle head, bordered in black; includes black "eye."
Habitat	Open, arid habitats including scrub oak barrens, deserts, and sparse woodlands.
Range	Primarily from Florida to California; may spread north to New England, plains states, and central California. California Dogface from Coast Ranges to Baja.
Life Cycle	Caterpillars feed on legumes, including clovers (*Trifolium* spp.). Adults, 2 flights.

Cloudless Sulfur *Phoebis sennae*

Although many of the "whys" of migration have yet to be answered, the annual movements of migratory butterflies such as Monarchs and Cloudless Sulfurs have long been an interesting puzzle. Such movements are, most likely, an adaptation to the tropical cycle of wet and dry seasons. The Monarch *(Danaus plexippus)* is a two-way migrant—some individuals make a round trip. The Cloudless Sulfur accomplishes a similar mission using two or more broods: some spring individuals move north, while late-summer butterflies head south.

Identification Wingspan 2¼–2¾". Overall yellow. Above: males clear yellow, females deep yellow or white with dark marginal wing spots and dark spot in center of FW.

Habitat Open areas including roadsides, beaches, and fields.

Range Florida west along Gulf Coast, at times as far as S. California; north to Canadian border.

Life Cycle Eggs cream-colored, turning reddish. Caterpillars' host plants mainly sennas *(Cassia* spp.). Adults, 2 flights in the North, all year farther south.

Little Yellow *Eurema lisa*

Several groups of butterflies, including the swallowtails, sulfurs, and blues, are often found in aggregations around damp or muddy areas. These so-called "mud puddling clubs" are usually made up only of males, who may be ingesting salts. The 18th-century lepidopterist John Abbot, observing this behavior in Little Yellows, reported that the butterfly "settles so many together at times to suck moist places that I have seen twenty in the compass of a hat."

Identification	Wingspan 1–1½″. Overall yellow with black. Above: bright yellow with black FW tips and borders; blackish areas often reduced in females. Below with some mottling and reddish spot on HW.
Habitat	Open areas; meadows, fields, and waste places.
Range	Southeastern states to S. New England and Great Lakes region; at times, to Arizona.
Life Cycle	Caterpillars feed on legumes. Adults, many flights; year-round in southern portion of range. Sporadic, but spectacular, fall flights to Bermuda and the Caribbean.

Harvester *Feniseca tarquinius*

The Harvester is the only North American representative of an unusual Old World group occurring mainly in the Asian and African tropics. It is also the only North American species whose caterpillars are carnivorous—they feed on aphids or plantlice. Adult Harvesters are usually also found near aphid populations. The tiny insects produce and excrete a sugary substance called "honeydew," which the butterflies ingest.

Identification	Wingspan 1⅛–1¼". Overall orange and black. Above: orange with dark borders and splotches. Below: HW with many oval- and crescent-shaped, gray vermiculations.
Habitat	Wetland edges, damp glades, and swampy areas; often associated with alders (*Alnus* spp.).
Range	E. United States and Maritime Provinces to Florida and Gulf Coast; west to E. Manitoba and central Texas.
Life Cycle	Short life cycle (3 weeks). Eggs greenish white. Caterpillars feed on aphids. Adults, 2 flights.

Little Copper *Lycaena phlaeas*

Certain plants and animals seem to thrive in areas disturbed by human activities or natural events. The Little Copper is right at home in waste places and disturbed sites. Despite this butterfly's choice of habitat and its tiny size, it is dazzling to see. The fiery tones, which lend a sparkling brightness to this and other coppers, owe their brilliance to the light-refracting scales.

Identification	Wingspan ⅞–1⅛″. Above: FW red-orange with dark splotches and borders; HW dark, with submarginal orange band. Below: gray with black markings; FW with some orange, HW with red-orange band.
Habitat	Disturbed areas—old lots, roadsides, and fields; above tree line in mountains.
Range	Throughout Northeast, south to Georgia, and west to Dakotas. Disjunct populations in mountains of Alaska, central California, and Oregon.
Life Cycle	Host plant is sheep sorrel *(Rumex acetosella)*. Mountain form, 1 flight; lowland form, 2 or more flights.

Purplish Copper *Lycaena helloides*

A species in which the male and female look quite different is said to exhibit sexual dimorphism. The Purplish Copper is a very good example; the dorsal wing surfaces of the male are quite different from the female's. Sexual dimorphism is also seen in the similar and closely related Dorcas Copper *(L. dorcas)*, which replaces the Purplish Copper in parts of the East.

Identification — Wingspan 1–1¼″. Above: male dull copper with iridescent purplish highlights; black spots and red crescents on border of HW; female orangy brown, more heavily spotted; red crescents on HW border. Below: both sexes with spotted, ochre FW and spotted, grayish-tan HW; reddish crescents near trailing edge.

Habitat — Mainly lowland areas, including weed fields and roadsides; sometimes foothills and mountain meadows.

Range — Great Lakes to S. British Columbia; south to Mexico.

Life Cycle — Caterpillars feed mainly on dock *(Rumex)* and cinquefoil *(Potentilla)*. Adults, several flights.

Coral Hairstreak *Satyrium titus*

Butterflies get their names from a variety of sources—some purely scientific, others fanciful. The generic name *Satyrium*, for example, comes from the Greek *Satyros*, the name for a group of diminutive woodland deities. (Some experts place this species in the genus *Harkenclenus*, named to honor Harry Kendon Clench, a noted hairstreak specialist.) The specific epithet, *"titus,"* refers to the Roman emperor. Shown at left is the California Hairstreak *(S. californica)*, which occurs only in the West.

Identification	Wingspan 1–1¼". Overall brown. Below: dark with white-edged spots on both wings (except in West); HW with bright coral markings along entire submargin.
Habitat	Open areas: fields, scrub woodlands, and canyons.
Range	Most of United States, except Florida, the extreme S. United States, the Southwest.
Life Cycle	Caterpillars feed on wild plums and wild cherries *(Prunus)* and serviceberry *(Amelanchier alnifolia)*.

Banded Hairstreak *Satyrium calanus*

In general, male butterflies use two methods—patrolling and perching—to canvass mating grounds for a suitable female. The male Banded Hairstreak uses the perching method; it can often be seen flying out from its favorite perch to inspect other insects, birds, even humans—and, of course, other Banded Hairstreaks. A female Banded Hairstreak is shown at left.

Identification | Wingspan 1–1¼". Above: dark brown to blackish with short, threadlike tails. Below: brownish with various light and dark lines; FW markings on mid-band continuous, forming unbroken rectangle; HW with small red chevron(s) and blue patches on trailing edge.

Habitat | Woodland edges, openings, and canyons.

Range | E. United States except N. Maine and S. Florida, west through Dakotas and central Texas; S. Wyoming south to New Mexico.

Life Cycle | Caterpillars feed on oaks (*Quercus* spp.) and nut trees (*Juglans* spp.). Adults, 1 flight per year; take nectar from a variety of blossoms.

Olive Hairstreak *Mitoura grynea*

During the second half of the 19th century, many farms in the Northeast were abandoned as more fertile and larger tracts opened in the Midwest. The fallow fields and meadows soon began the process of natural succession. One of the early changes was the growth of eastern redcedars *(Juniperus virginiana)* on the abandoned acreage—a boon to the Olive Hairstreak, whose life cycle centers about this tree. As this habitat continues to change, a decline in Olive Hairstreaks may be expected. Also known as *Mitoura gryneus.*

Identification	Wingspan ⅞–1″. Above: brownish to orange. Below: bright green; FW with band of white marks, HW with scattered white chevrons or crescents. Similar Hessel's Hairstreak *(M. hesseli)* has white dot on FW below; occurs with Atlantic white-cedar *(Chamaecyparis thyoides)*.
Habitat	Old fields and hillsides with red cedar.
Range	Dakotas to New England, Texas, and Florida.
Life Cycle	Larvae feed on cedars *(Juniperus)*. Adults, 1–2 flights.

Brown Elfin *Incisalia augustinus*

The nine species of North American elfins (*Incisalia* spp.) are harbingers of spring—they overwinter as chrysalises, emerging in early spring to make their single annual flight. Because elfins are small, inconspicuous, and somewhat drab, they are passed over by some as the LBJs ("Little Brown Jobs") of the butterfly world. Actually, the muted browns and grays of each species have subtle but characteristic patterns and hues. The Hoary Elfin *(I. polios)* is shown at left.

Identification | Wingspan ¾–1⅛". Brown overall, without the short tails that occur in some elfin species. Below: brownish to rust, may be mahogany tone to outer portion of HW.

Habitat | Blueberry barrens, woodland edges, and chaparral.

Range | Appalachians north to Maritime Provinces, west through N. United States and central Canada to Alaska, and south to Baja California.

Life Cycle | A wide variety of host plants includes blueberries (*Vaccinium* spp.), wild lilac *(Ceanothus)*, and dodder *(Cuscuta)*.

64

Gray Hairstreak *Strymon melinus*

The male Gray Hairstreak, like the Banded Hairstreak, is a "perching" butterfly. While the life cycle of many butterflies is closely tied to a single species of plant, the Gray Hairstreak is unusual in the number of host plants it finds suitable. Caterpillars of this species eat dozens of plants from numerous plant families, including legumes (Leguminosae), mallows (Malvaceae), snapdragons (Scrophulariaceae), mints (Lamiaceae), and cactus (Cactaceae).

Identification	Wingspan 1–1¼″. Above: dark gray to brownish with orange spot next to short tail. Below: pale gray with thin black-and-white bands variably edged in red; orange and blue markings and terminal black spot on trailing tip of HW.
Habitat	Fields with some second growth shrubs, open woodlands, disturbed areas.
Range	S. Canada throughout United States.
Life Cycle	Caterpillars feed on a wide variety of host plants. Adults, 2 to many flights.

Eastern Tailed Blue *Everes comyntas*

This butterfly and its close relative, the Western Tailed Blue *(E. amyntula)*, look very much alike. The latter, however, is slightly larger and paler, and has fewer and lighter spots below. The Western Tailed Blue generally replaces the Eastern west of the Rocky Mountains. There is some crossover—a few isolated populations of the Eastern Tailed Blue occur in California, Oregon, Arizona, and New Mexico; the Western Tailed Blue occurs in the northern portions of Minnesota, Michigan, and the Gaspé Peninsula in Quebec.

Identification	Wingspan ¾–1″. Above: blue to gray with tiny, threadlike tails. Below: ash-gray with black spots on both wings and orange patch on HW margin.
Habitat	Open habitats: roadsides, waste places, and fields.
Range	E. United States except N. Minnesota and Maine, and S. Florida; S. New Mexico, Arizona, and lowland California and Oregon.
Life Cycle	Caterpillars feed on legumes (Leguminosae). Adults, 2–5 flights.

68

Spring Azure *Celastrina argiolus*

For the professional lepidopterist, the Spring Azure is a complex butterfly; the species, as currently defined, includes at least four subspecies and six forms. All are very similar. Scientists pursuing the matter have recently separated out the Sooty Azure, which was previously thought to be a form; it is now considered a species, *C. ebenina*. The observations made by amateur butterfly enthusiasts—especially with regard to local and seasonal varieties of the Spring Azure—could prove to be of real value in this ticklish taxonomical puzzle. Also known as *C. ladon*.

Identification	Wingspan ¾–1¼". Above: deep blue to gray; no tails or orange spots. Below: gray with charcoal spots. Variable.
Habitat	Wooded edges, brushy fields, and wetland margins.
Range	Alaska, Canada, and United States. Absent from portions of Florida, Louisiana, and Texas.
Life Cycle	Host plants include dogwoods *(Cornus)* and blueberries *(Vaccinium)*.

Silvery Blue *Glaucopsyche lygdamus*

The Silvery Blue is a pretty, slow-flying species with much geographical variation. It is one of a small number of *Glaucopsyche* species in North America; its most famous relative is the Xerces Blue *(G. xerces)*, now extinct. The last of these butterflies was observed in 1934 in San Francisco; its habitat was destroyed by urban expansion.

Identification	Wingspan 1–1¼". Variable. Above: males iridescent silvery blue with narrow black border, females dark blue to brown with wider border. Below: gray with single row of white-rimmed, black dots.
Habitat	Open woodlands, brushy fields, mountain meadows, and canyons.
Range	Widespread but often local. W. United States from Alaska to Baja California east to Arizona, New Mexico, and Oklahoma; also locally in N. Alabama and Georgia north to West Virginia and Pennsylvania.
Life Cycle	Caterpillars feed on legumes (Leguminosae). Adults, 1 flight; early spring to mid-July, depending on climate.

72

Mormon Metalmark *Apodemia mormo*

Some butterfly taxonomists consider the metalmarks to be in their own family (Riodinidae), while others classify them as a subfamily (Riodininae) within the Gossamer Wings. Whatever their ultimate disposition, this largely tropical group is represented by about 20 species in North America; most are found in the southern United States. The group's name refers to the metallic markings that often adorn these butterflies. The Mormon Metalmark lacks these metallic highlights but does have brightly patterned wing surfaces.

Identification Wingspan ¾–1¼". Overall black with bold white markings. Both surfaces of inner FW have rust-colored patch, often with similar color on upper HW; HW below is pearl gray with prominent white patches.

Habitat Arid dunes, slopes, canyons, and chapparal.

Range Northwest, S. British Columbia, and South Dakota south through central Rocky Mountains to W. Texas.

Life Cycle Caterpillars feed on wild buckwheats (*Eriogonum* spp.). Adults, 1–3 flights in spring and late summer.

74

American Snout *Libytheana carinenta*

Snout butterflies (Libytheidae), which form a small family of ten or so species worldwide, get their name from their very pronounced facial appendages, or palpi (singular, palpus). The function of the snout is not known with certainty, but one researcher suggests that this is a camouflage adaptation: when the butterfly is at rest, the wings resemble a leaf and the snout looks like a leaf stem. Until recently there was believed to be another distinct species of snout, *L. bachmanii;* the two are now considered to be the same species.

Identification	Wingspan 1⅝–1⅞″. Extended palpi suggest a snout. FW above orange and black with bold white patches; tips squared.
Habitat	Found in a variety of locales, including wood edges, forests, canyons, and hillside scrub.
Range	S. United States; emigrates northward to New England west through the Dakotas (makes no return trip).
Life Cycle	Caterpillars feed on hackberry (*Celtis* spp.). Adults fly year-round in South.

Gulf Fritillary *Agraulis vanillae*

The Gulf Fritillary is one of several longwings occurring in North America. Although this species is superficially similar to the true fritillaries, some taxonomists do not consider the relationship to be a close one. Longwing caterpillars feed on passion vines (*Passiflora* spp.), making both the larvae and adult butterflies toxic to many predators. Female Gulf Fritillaries have an organ at the tip of their abdomen called a "stink club," which emits strong scents (pheromones) that either attract or repel males, depending on the female's receptive state.

Identification	Wingspan 2½–2⅞". Above: deep orange-red with tiny white spots on FW and black spots scattered over both wing surfaces. Below: brown to orange with prominent metallic silver markings on both wings.
Habitat	Subtropical woodlands, fields, roadsides, and parks.
Range	Resident in S. Florida, Texas, Arizona, and California; emigrates to Long Island, Wisconsin, and Colorado.
Life Cycle	Caterpillars feed on passion flowers (*Passiflora* spp.). Adults, several flights.

Great Spangled Fritillary *Speyeria cybele*

As is true of many fritillaries, the Great Spangled Fritillary's life cycle is closely tied to violets (Violaceae). In late spring, males emerge from the pupal stage; the females begin to fly about a month later, and then mating takes place. Many females survive until late summer, when they lay eggs on or near violet host plants. Adults take nectar from black-eyed susans, thistles, and other flowers.

Identification   Wingspan 2⅛–3″. Above: orange with numerous black lines, spots, and chevrons. Below: HW with many metallic silver spots, most forming 2 rows enclosing wide yellow band. Similar Aphrodite Fritillary *(S. aphrodite)* has narrow yellow band below and black dot on basal portion of upper FW.

Habitat   Wet meadows and open woodlands.

Range   Georgia to Arkansas and Oklahoma to central California; north to S. Canada.

Life Cycle   Eggs hatch in fall. Caterpillar overwinters, emerges in spring to feed on violets (*Viola* spp.). Adults, 1 flight.

Atlantis Fritillary *Speyeria atlantis*

The Atlantis and other fritillaries give off scents (pheromones) to attract mates. The male's pheromones are located in specialized scales on the forewing veins. During his courtship dance, the male flicks his wings and disperses a scent that is picked up by the female's antennae. These "perfumes" may counteract the visual similarity of many of the fritillary species, by helping the butterflies to attract mates of their own species.

Identification	Wingspan 1¾–2⅝". Regionally variable. Above: orange with numerous black lines, spots, and chevrons; black wing margins solid and wide in East. Below: HW has many silver spots on HW in East; no silver spots in parts of the West; opaque spots in Rocky Mountains. All forms have narrow yellow submarginal band.
Habitat	Woodland openings and trails, meadows, and prairies.
Range	Nova Scotia to Alaska south to California, New Mexico, Great Lakes region, and Virginia.
Life Cycle	Caterpillars feed on violets (Violaceae). Adults, 1 flight midsummer.

Silver-bordered Fritillary *Boloria selene*

The range of the Silver-bordered Fritillary is described by biologists as being "holarctic," indicating that Silver-bordered Fritillaries can be found around the globe in a region that includes parts of North America, Europe, and Asia. The Silver-bordered Fritillary and the Meadow Fritillary *(B. bellona)* are among the lesser fritillaries. Normally they lack the silver spots of the greater fritillaries; however, the Silver-bordered Fritillary is the exception to the rule.

Identification	Wingspan 1⅜–2″. Above: orange with numerous black lines, spots, and chevrons. Below: HW has several rows of silver spots and one row of dark, submarginal spots.
Habitat	Wet meadows and bogs, often with some shrub growth.
Range	S. Alaska through Colorado and Iowa and east to Nova Scotia and mid-Atlantic states.
Life Cycle	Caterpillars feed on violets (Violaceae). Adults, 1–3 flights.

Meadow Fritillary *Boloria bellona*

The Meadow Fritillary, a typical lesser fritillary, lacks silver spots. Like most of its relatives, it dwells chiefly in northern areas, where it is frequently seen near bogs. This species and the Silver-bordered Fritillary (*B. selene*), however, range farther south than most lesser fritillaries. Some experts have classified both species in the genus *Clossiana*.

Identification Wingspan 1¼–1⅞″. FW tips appear clipped. Above: orange with numerous black lines, spots, and chevrons. Below: no silver spots; FW with black markings, HW orange-brown with outer portion having lavender-gray cast.

Habitat Wet meadows and bogs.

Range E. Quebec to North Carolina and northwest through midwestern states, North Dakota, Colorado, and Washington. Most of S. Canada.

Life Cycle Caterpillars feed on violets (Violaceae). Adults, 1–3 flights.

Variegated Fritillary *Euptoieta claudia*

The chrysalises of many butterflies are distinctly unappealing to look at. Notable exceptions are the pupae of the Monarch *(Danaus plexippus)* and the Variegated Fritillary. The waxy green chrysalis of the Monarch is bejeweled with flecks of gold. The chrysalis of the Variegated Fritillary, a luminescent bluish white dotted with gold, orange, and black markings, is equally stunning.

Identification Wingspan 1¾–2¼". Above: orange with numerous black lines, spots, and chevrons. Below: lacking silver spots but with large areas of whitish brown on both wings. HW with marbled pattern.

Habitat Open habitats including fields, meadows, waste places, and fallow fields.

Range Arizona to Gulf Coast states; emigrates north to S. Canada.

Life Cycle Caterpillars feed on a wide variety of host plants including violets (*Viola* spp.) and passion flowers (*Passiflora* spp.). Adults fly all year in South.

88

Silvery Checkerspot *Chlosyne nycteis*

Butterflies pass through four stages during their life cycle, allowing them to adapt to a wide range of climates and habitats. In North America, butterflies occur in such diverse areas as the tundra, the arctic-alpine zone of treeless mountaintops; the subtropical forest; and the desert. These tough little insects survive annual periods of severe cold or drought in a resting state known as diapause. The Silvery Checkerspot, sometimes called Silvery Crescentspot or *Charidryas nycteis*, overwinters as a caterpillar in the North.

Identification	Wingspan 1⅜–1¾″. Above: dark brown to black with some orange; HW with marginal black dots, some white-centered. Below: HW with 2 areas of silvery white markings.
Habitat	Fields, streamsides, and moist meadows.
Range	E. North America from S. Canada to Georgia and central Texas; Colorado, New Mexico, and Arizona.
Life Cycle	Caterpillars feed on a variety of daisy family members (Compositae).

90

Pearl Crescent *Phyciodes tharos*

This species, also called the Pearly Crescentspot, is closely related to the Northern Pearl Crescent *(P. selenis)*, a newly recognized species that was previously considered part of *P. tharos*. Although the differences between the two species are complex rather than simple, the more northerly *selenis* is larger. In the East, *tharos* males often have black-and-white antennae clubs, while *selenis* males have black-and-orange clubs. Some hybridization apparently occurs between various populations of these two species.

Identification	Wingspan 1–1½". Above: orange with black borders, scattered black markings. Below: HW yellow with brownish vermiculations; dark patch with crescent.
Habitat	A variety of open areas, including meadows, fields, and roadsides.
Range	S. Alberta east to New England and south to Florida, Gulf Coast, Texas, and S. Arizona.
Life Cycle	Caterpillars feed on asters (*Aster* spp.). Adults, 2 flights; all year in southern portions of range.

92

Field Crescent *Phyciodes pratensis*

The Field Crescent is the most commonly encountered crescentspot of western mountains. Formerly called *Phyciodes campestris*, and also known as the Field Crescentspot, it is at home in a wide range of climates and habitats. It flies in mountain meadows and northerly prairies from the Northwest Territories to southern California, Arizona, and New Mexico.

Identification	Wingspan 1–1⅜″. Above: dark; numerous yellow and orange markings form checkered pattern; less orange above than Pearl Crescent. Below: both wings largely orange to cream; sparse black markings on FW; HW with crescent as in Pearl Crescent *(P. tharos)*.
Habitat	Tundra, mountain meadows, fields, and plains.
Range	West of Rocky Mountains from N. Canada to southwestern states.
Life Cycle	Caterpillars feed on asters (*Aster* spp.). Adults, 1–3 flights depending on climate and elevation.

Baltimore *Euphydryas phaeton*

The Baltimore, one of a group of butterflies known as the checkerspots, has bright reddish-orange markings that serve as a warning to predators that this insect is toxic. And indeed it is, sometimes. Baltimores that feed on turtlehead *(Chelone glabra)* cause gastric upset in the birds that eat them. Individual Baltimores that eat more benign host plants are apparently more palatable to predators.

Identification	Wingspan 1⅝–2½". Overall: black with reddish-orange and whitish checks.
Habitat	Wet meadows and arid hillsides.
Range	New England to SE. Manitoba and south to Arkansas and Georgia.
Life Cycle	Caterpillars feed on a variety of host plants including turtlehead *(Chelone glabra)* and plantain *(Plantago lanceolata)*. Larvae choose certain host plants early in their development and others later on. Adults, 1 flight from May–July.

Anicia Checkerspot *Euphydryas chalcedona*

This common western checkerspot is known for its
variability. Eight subspecies are recognized in
California, and at least two dozen other varieties have
been described. The typical form has a black-and-
cream-colored checkered pattern and red wing border.
Common variants include those with extensive areas of
brick red, white, orange, or yellow on the dorsal
forewing. This species is relatively tame, and can often
be approached for close observation and photography.
Also known as the Chalcedon Checkerspot.

Identification	Wingspan 1⅜–2″. Variable. Above: black-and-cream-checkered pattern and thin brick-red wing border. Below: HW with red and cream bands.
Habitat	Arid hillsides, chaparral, open woodlands, and tundra.
Range	S. Alaska through California; east to New Mexico, Colorado, and Montana.
Life Cycle	Caterpillar host plants include figworts (Scrophulariaceae) and snowberry (*Symphoricarpos albus*). Adults, 1 flight.

Question Mark *Polygonia interrogationis*

The *Polygonia* butterflies are called, familiarly, anglewings—for the shape of their wings. The three anglewings described in this guide overwinter as adults —an unusual trait in butterflies. Woodpiles, crevices behind loose bark, and outbuildings all provide suitably sheltered spots for hibernating, but some individuals migrate south. Relatively few flowers are in bloom when adults fly in early spring and late summer, but the anglewings find nourishment in tree sap and rotting fruit.

Identification	Wingspan 2⅜–2⅝″. Above: orange with black on HW (heaviest in summer) and black markings on FW. Below: marbled or plain brownish to violet; HW with 2-part silvery "question mark." Tails on HW.
Habitat	Woodland openings, paths, and brooks.
Range	E. United States to Rocky Mountains; S. Canada.
Life Cycle	Caterpillars feed on elms (*Ulmus* spp.), hops (*Humulus* spp.), and hackberry (*Celtis* spp.). Adults overwinter; 2–3 flights in early spring and late summer.

100

Comma *Polygonia comma*

Butterflies, like most animals, face the dual challenges of attracting a mate and avoiding predators. One evolutionary solution to these potentially conflicting aims is apparent in the coloration and patterning of the wing surfaces of some butterflies. The Comma's upper wing has bright orange patches, which presumably are attractive to potential mates. In contrast, the ventral surfaces of the wings show cryptic, brownish barklike patterns. Also called the Hop Merchant.

Identification	Wingspan 1¾–2″. Above: orange with black spotting and borders. Below: brownish; HW with silvery "comma" with hooked ends. Violet highlights at times. Similar to Question Mark but smaller, with stubbier tails.
Habitat	Woodland openings, paths, and brooks.
Range	S. Canada and E. United States to E. Texas, central Colorado, and Dakotas; absent from most of Florida.
Life Cycle	Host plants mainly hops *(Humulus lupulus)* and nettles *(Urtica, Boehmeria* spp.). Adults, 2–3 flights.

102

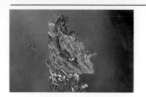

Satyr *Polygonia satyrus*

Also known as the Satyr Anglewing, this species is the common *Polygonia* of much of the West, replacing the Comma there. The butterfly was first described in 1869 by William Henry Edwards, whose *Butterflies of North America* is one of the cornerstones of North American butterfly studies. His memory is recalled in the names of several species, including Edwards' Hairstreak *(Satyrium edwardsii)*.

Identification	Wingspan 1¾–2″. Above: lighter than other anglewings, golden orange (darker in Far West) with scattered black markings. Below: brown to tan marbled pattern or plain; HW with silvery comma.
Habitat	Wooded areas including foothills, canyons, and parks.
Range	British Columbia to Arizona; east across S. Canada. Scattered populations in N. Maine, Michigan, Wisconsin, and Minnesota.
Life Cycle	Host plants largely nettles (*Urtica* spp.). Adults, 2–3 flights.

104

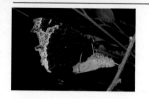

Mourning Cloak *Nymphalis antiopa*

A tradition of folklore seems to attend most animal groups, and butterflies are no exception. Thaddeus W. Harris, a 19th-century entomologist, recounted that Mourning Cloak caterpillars were believed "venomous, and capable of inflicting dangerous wounds." Indeed, they look ferocious and are covered with protective spines and hairs; but this awe-inspiring creature is benign, and in time transforms itself into one of our most dramatic and impressive butterflies.

Identification   Wingspan 2⅞–3⅜". Above: deep maroon with marginal row of violet-blue marks and yellow borders. Below: dark gray.

Habitat   Woodland openings and edges; streamsides; gardens.

Range   Most of North America; absent from Far North and subtropics; rare in Deep South.

Life Cycle   Caterpillars feed on willows (*Salix* spp.), elms (*Ulmus* spp.), and poplars (*Populus* spp.). Adults, normally 1 generation, hibernate; sometimes fly during midwinter thaws. Adults may live up to 11 months.

106

Milbert's Tortoiseshell *Nymphalis milberti*

Four tortoiseshell butterflies occur in North America: Milbert's, Compton *(N. vau-album)*, and California *(N. californica)* tortoiseshells and the Mourning Cloak *(N. antiopa)*. Like their close relatives, the anglewings, the adult tortoiseshells hibernate. Milbert's Tortoiseshell, seen here, is regularly found farther north than the others. Its relative abundance in any given locale may fluctuate radically from year to year. This species was formerly called *Aglais milberti*.

Identification	Wingspan 1¾–2″. Above: bright, 2-toned, orange-and-yellow band borders dark brown areas; 2 orange patches on each FW. Below: largely dark brown with lighter, wide band.
Habitat	Mountaintops to dry desert washes; most places in between.
Range	S. Alaska to Pacific Northwest (rarely to S. California) and east to Newfoundland.
Life Cycle	Caterpillars feed on nettles (*Urtica* spp.). Adults, 2, perhaps 3 flights.

Red Admiral *Vanessa atalanta*

Although it is often difficult to become acquainted with individual butterflies, the Red Admiral can be an exception. Because of this species' territorial behavior, a Red Admiral can often be found in the same place for several days or even weeks. This butterfly is particularly fond of sunlit gardens and paths, and once you have discovered a Red Admiral's territorial perch you will have a good subject for observation.

Identification
Wingspan 1¾–2¼". Above: black with orange-red bar across middle of FW and HW border. Below: black and brown marbling with pink bar on FW.

Habitat
Woodland openings, gardens, meadows, streamsides, and savannahs.

Range
Throughout North America except northernmost Canada and Alaska.

Life Cycle
Caterpillars feed on nettles *(Urtica, Boehmeria)* often in leaf nests. Adults hibernate and northern areas are recolonized each year by migrants. Normally 2 flights per year.

110

American Painted Lady *Vanessa virginiensis*

The common names for some butterflies have gone through a bewildering number of changes. This species has variously been known as the Painted Beauty, Hunter's Butterfly, Hunter's Cynthia, Marbled Cynthia, and Scarce Painted Lady; perhaps its striking color, and the marked difference between the upperside and underside, account for this range of evocative epithets.

Identification	Wingspan 1¾–2⅛". Above: orange with black markings; FW with white spots in black tip; HW with small, bluish spots circled by black. Below: FW with large, pink area; HW, 2 large eyespots near border.
Habitat	Open areas with wildflowers; fields, meadows, and streamsides.
Range	S. Canada through United States and Mexico. More common in East; scarce in West.
Life Cycle	Caterpillars feed on members of the daisy family (Compositae). Adults, 2–3 flights. Numbers fluctuate widely from year to year.

112

Painted Lady *Vanessa cardui*

This species has been called the Thistle Butterfly; in fact, the *Vanessa* species as a group are referred to as thistle butterflies because they prefer thistles as a nectaring source. The Painted Lady is also called the Cosmopolite, a reference to its worldwide distribution. Indeed, it is the most widespread butterfly in the world, found on every continent except Antarctica—although its appearances in Australia are rare.

Identification	Wingspan 2–2¼". Above: orange with black markings; FW with white spots in black tip; HW with small black or black-rimmed bluish spots near border. Below: FW with large pink area; HW with row of small black eyespots near border.
Habitat	Found in almost any setting.
Range	Subarctic North America to central America.
Life Cycle	Caterpillars feed on daisy family members (Compositae), especially thistles (*Cirsium* spp.). Individuals from southern populations move northward each year to recolonize colder areas.

114

West Coast Lady *Vanessa annabella*

This species provides an interesting example of the workings of zoological nomenclature. For well over a century and a half, this butterfly carried the Latin name *V. carye* and was thought to be identical with a South American species of the same name. In the early 1970s, however, the West Coast Lady was redescribed. Because every species must have a unique scientific name, we now know this one as *Vanessa annabella*.

Identification	Wingspan 1¾–2″. Above: orange and black with orange bar in black area of FW tip; HW lower margin with blue eyespots. Below: FW with corresponding orange or yellow bar; also with pink patch. HW marbled tans and yellows. FW apex somewhat squared.
Habitat	Lowland meadows, hillsides, gardens, and waste areas.
Range	Mainly west of Rocky Mountains from British Columbia to Baja California; east to E. Colorado and W. Kansas.
Life Cycle	Caterpillars feed on mallows (Malvaceae). Adults fly all year in warmer parts of range.

116

Buckeye *Junonia coenia*

The Buckeye lives year-round in the southern United States, but farther north the adults cannot overwinter. The seasonal changes taking place at summer's end prompt at least some of the northern Buckeyes to move south. This phenomenon can be quite impressive along the New England and mid-Atlantic coasts.

Identification   Wingspan 2–2½". Above: brownish with 2 orange bars on FW and 2 different-sized eyespots on each wing. Below: brownish or red-pink (winter brood) with corresponding orange bars; eyespots from above less prominent. Similar Mangrove Buckeye *(J. evarete)* of south Florida has 2 eyespots of same size on HW above.

Habitat   Open fields and meadows, railroad tracks, and waste places; coastal beaches in late summer and fall.

Range   S. United States; north to S. Canada in summer.

Life Cycle   Caterpillars feed on plantains (Plantaginaceae), vervains (Verbenaceae), and snapdragons (Scrophulariaceae). Adults fly all year in South; 2 flights northward.

118

White Admiral *Limenitis arthemis arthemis*

A very early American account of the White Admiral was written by Thomas Say, one of the greatest American naturalists. In the early 19th century, he wrote *American Entomology*, a pioneering work on New World insects. In his account of the White Admiral, Say notes that he had collected this butterfly along the Arkansas River—a location that seems improbable today for this northerly species. Sometimes called the Banded Purple, it was formerly placed in the genus *Basilarchia;* it is now considered conspecific with the Red-spotted Purple *(L. a. astyanax).*

Identification   Wingspan 2⅞–3⅛". Above: black with white band through both wings. Below: dark brown with corresponding white band and submarginal row of brick-red spots.

Habitat   Open hardwood forests and woodland edges.

Range   Alaska to New England.

Life Cycle   Caterpillars feed on birches (*Betula*), willows (*Salix*), and poplars (*Populus*). Adults, 1–2 flights.

120

Red-spotted Purple *Limenitis arthemis astyanax*

Most authorities consider that the Red-spotted Purple is conspecific with the White Admiral, the former being a southerly subspecies, the latter northerly. The southern subspecies, with its largely black and iridescent blue upper wing surfaces, is considered a mimic of the Pipevine Swallowtail, which is toxic to predators. Because pipevines are not found in the White Admiral's range, such mimicry would have no selective advantage—and, indeed, it does not take place.

Identification Wingspan 3–3⅜". Above: black with iridescent blue at borders and several white spots at FW tips. Below: brownish with submarginal and basal brick-red marks.

Habitat Open hardwood forests, woodland edges; also washes and canyons in the Southwest.

Range New England to the Dakotas and to the Gulf states; also Arizona and New Mexico.

Life Cycle Host plants are willows (*Salix*), poplars (*Populus*), and wild cherries (*Prunus*). Adults, 1–3 flights.

Viceroy *Limenitis archippus*

The Viceroy is a classic mimic. In the North, where populations of Monarchs *(Danaus plexippus)* are plentiful, the Viceroy has evolved to resemble this species. Predators learn to avoid not only the toxic Monarch but the look-alike Viceroy as well. In the southern United States, Monarchs are scarce; but the Viceroy of the South has evolved to resemble the toxic Queen *(D. gilippus)*.

Identification | Wingspan 2⅝–3″. Above and below: orange with white-spotted black wing borders and black venation. HW with lateral black line perpendicular to veins. Background color ranges from light orange to mahogany, depending on local model for mimicry.

Habitat | Meadows, fields, and wetland edges.

Range | British Columbia and Manitoba south to Great Basin and east to Newfoundland. South to Gulf Coast and parts of the Southwest.

Life Cycle | Caterpillars feed on willows *(Salix* spp.) and poplars *(Populus* spp.). Adults, 2–3 flights.

124

Weidemeyer's Admiral *Limenitis weidemeyerii*

The Weidemeyer's Admiral is found primarily in the Rocky Mountains, while the similar Lorquin's Admiral (*L. lorquini*) occurs farther west, from W. Montana to the Pacific Coast. Lorquin's is distinguished from the Weidemeyer's by its brick-red forewing tips. Both of these admirals, as well as other species of limenitines, have a characteristic sailing aspect to their flight.

Identification Wingspan 2¾–3⅜″. Above: black with broad, pure white band crossing both wings. Below: significant amounts of white on both wings; FW with reddish bars, HW with transverse bar of reddish spots and blue crescents.

Habitat Streamsides lined with willows and cottonwoods. To 11,000′ on mountainsides.

Range S. Alberta through New Mexico and Arizona in Rocky Mountain states; also Nevada to east-central California.

Life Cycle Caterpillars feed on willows (*Salix* spp.) and poplars (*Populus* spp.). Adults, normally 1 flight.

California Sister *Limenitis bredowii*

At first glance, the California Sister seems quite different from the other limenitines. The white wing bands, bright forewing tip markings, and the pattern on the ventral hindwing, however, are reminiscent of the various admirals. This is another of the "mud puddling" butterflies, and congregations of male California Sisters are often found at damp, streamside patches. California Sisters are sometimes placed in the genus *Adelpha*.

Identification	Wingspan 2⅞–3⅜″. Above: brownish black with white band and bright orange patch near FW tip. Below: similar, with submarginal and basal blue markings.
Habitat	Oak woodlands and riparian canyons; also coastal areas and California Channel Islands.
Range	S. Washington to Mexico, Colorado, and central Texas.
Life Cycle	Caterpillars feed on oaks (*Quercus* spp.). Adults, 1–2 flights.

Hackberry Butterfly *Asterocampa celtis*

Asterocampa means "star caterpillar," probably in reference to the starlike horns on caterpillars of the Hackberry. The adults of this woodland species are sexually dimorphic; with females larger and paler than the males. Although there is also considerable geographic variation in the Hackberry Butterfly, most experts believe that all forms belong to a single species.

Identification   Wingspan 1¾–2¼″. Variable. Above: Tan to tawny with numerous black markings; FW with 1 large, black eyespot and white markings in black wing tip. Below: marbled, lavender to tan; numerous eyespots.

Habitat   Deciduous woodlands, paths, canyons, and edges—invariably associated with hackberry (*Celtis* spp.).

Range   S. New England to Colorado, Arizona, and the Gulf Coast.

Life Cycle   Caterpillars feed on hackberries (*Celtis* spp.). Adults, 1–2 flights.

Northern Pearly Eye *Enodia anthedon*

This is a northern, forest-dwelling satyr. Its close relative, the Pearly Eye (*E. portlandia*, shown at left), is a southern species that prefers areas with cane adjacent to swamps. While most butterflies are creatures of sunlight, the Northern Pearly Eye is right at home in the dappled shadows of forested areas. Even the courtship of this species takes place at twilight, as males sally forth from woodland perches in search of receptive females.

Identification	Wingspan 1⅝–2″. Above: tan with numerous eyespots; 5 on each HW. Below: brown, often with a lavender cast; eyespots have pupils. Similar Pearly Eye (*E. portlandia*) has orange antennal clubs; Northern Pearly Eye (*E. anthedon*) has black antennal clubs.
Habitat	Woodland edges and openings.
Range	S. Manitoba east to Newfoundland and south through the Appalachians and to Arkansas.
Life Cycle	Caterpillars feed on various grasses (Poaceae). Adults, 1 flight to the north, 2 to the south.

Appalachian Eyed Brown *Satyrodes appalachia*

This butterfly, sometimes called the Appalachian Brown, is often found near its sibling species, the Northern Eyed Brown *(S. eurydice)*. The two can, however, be distinguished by their habitat as well as physical characteristics. While Appalachian Eyed Brown likes wooded and shrubby swamps, the Northern Eyed Brown prefers open wetlands and marshes. The line on the ventral hindwing is sharply zigzagged (near the bottom of wing) on Northern Eyed Brown and wavy on Appalachian Eyed Brown.

Identification	Wingspan 1⅝–2″. Tan above and below with numerous eyespots adjacent to wing margins. Below: line closest to eyespots on HW wavy.
Habitat	Wooded or shrubby sedge swamps.
Range	S. Minnesota through New England and S. Quebec; south to N. Florida.
Life Cycle	Caterpillars feed on sedges (*Carex* spp.). Adults, 1 flight to the north, 2 to the south.

Little Wood Satyr *Megisto cymela*

While most species of butterflies decline in numbers or are locally extirpated as their habitat is altered, the Little Wood Satyr seems to have survived, even thrived, in moderately developed areas. Indeed, in midsummer this satyr can be one of the more abundant species—especially where forest fragmentation has occurred. Like the white-tailed deer and red fox, this butterfly has benefited from the increased edges created by woodland divisions.

Identification   Wingspan 1¾–1⅞". Above: brown with 2 eyespots on each wing. Below: tan corresponding eyespots and 2 dark, wavy lines crossing wings; smaller eyespots also apparent. Eyespots have double pupils.

Habitat   Woodland edges, openings, and thickets; also salt marshes and hammocks.

Range   E. Canada and E. United States except southernmost Florida; west to Dakotas, E. Colorado, and Texas.

Life Cycle   Caterpillars feed on grasses (Poaceae). Adults, 1 flight to the north, 2 to the south.

Plain Ringlet *Ceononympha tullia*

During the last quarter century, the Plain Ringlet has expanded its range into the northeast. Variously called Prairie Ringlet, Inornate Ringlet, California Ringlet, or just Ringlet (and sometimes *C. inornata*), this species shows wide geographical variation. Some authorities characterize these butterflies as the "*tullia* complex," indicating that classification questions remain unanswered. Fortunately for the butterfly enthusiast, ringlets are fairly easy to identify. A similar Ocher Ringlet *(C. ochracea)* is shown at left.

Identification	Wingspan 1–1⅞". Variable. Above: reddish brown to cream. Below: FW orange to gray, often with black, yellow-rimmed eyespot; HW tan to gray, with whitish wavy midline, sometimes with several submarginal eyespots. Weak, tilting flight.
Habitat	Grasslands, including prairies, meadows, and old fields.
Range	Alaska to Labrador, New England, Great Lakes, Rockies, California, and the Southwest.
Life Cycle	Host plants are grasses (Poaceae). Adults, 1–2 flights.

Common Wood Nymph *Cercyonis pegala*

Also called the Large Wood Nymph and the Blue-eyed Grayling, this butterfly is our largest and most abundant wood nymph. It is known for its variability. Geographic populations differ as to overall size, color, presence or absence of yellow forewing patch, and number and size of eyespots. This species is the only wood nymph regularly found east of the Mississippi.

Identification | Wingspan 2–2⅞". Variable. Overall tan to dark brown; often 2 FW eyespots in orange field. Sometimes numerous HW eyespots. Commonly occurring types include 1 form with bright orange FW patch and 1 without. A western variation has a yellow or white flush on the ventral FW and numerous ventral HW eyespots.

Habitat | Grassy borders of woodland edges and openings; also meadows, fields, and prairies.

Range | S. Canada and United States, absent from S. Florida and Southwest.

Life Cycle | Caterpillars feed on grasses (Poaceae). Adults, 1 flight.

140

Monarch *Danaus plexippus*

The Monarch is one of our best-known insects, and many people are familiar with the story of its metamorphosis and migration. So popular is this butterfly, in fact, that attempts are currently underway to designate it the national insect of the United States. At the same time various conservation groups, led by the Xerces Society, are engaged in efforts to preserve wintering areas of Monarchs in California and Mexico.

Identification — Wingspan 3½–4″. Above: orange with prominent black veins and borders. Male with oval-shaped scent patch on HW vein. Below: as above, only HW and part of the FW is distinctly paler. No transverse black line on HW.

Habitat — Open fields, meadows, prairies, and marshes, especially those with milkweeds (*Asclepias* spp.). Widespread during migrations.

Range — Most of North America from S. Canada south.

Life Cycle — Caterpillars feed on milkweeds and milkweed vines (*Asclepias* spp.). Adults may fly all year, depending on climate.

142

Queen *Danaus gilippus*

The Queen and the Monarch *(D. plexippus)* are our two most common milkweed butterflies. The Queen occurs in southern areas where there are few Monarchs. Like other milkweed butterflies, the male Queen has small brushes, or "hair pencils," on his abdomen. During courtship, the male transfers scent scales from a patch on his hindwing to these brushes and then in turn to the antennae of the female. Only when this rather complex process is complete is the female receptive to mating.

Identification   Wingspan 3–3⅜". Chestnut brown with darker borders and numerous white flecks on FW, at times on HW. No transverse black line on HW.

Habitat   Fields, meadows, dunes, and deserts with associated milkweeds *(Asclepias* spp.).

Range   S. Florida and Keys, Southwest; occasionally emigrates north to Nevada, Kansas, and North Carolina.

Life Cycle   Caterpillars feed on milkweeds and milkweed vines *(Asclepias* spp.). Adults fly all year in southern portion of their range.

144

Silver-spotted Skipper *Epargyreus clarus*

This common and relatively large skipper occurs regularly in areas associated with human activity. Gardens, parks, and suburban neighborhoods all provide welcome habitats for the Silver-spotted Skipper. This species makes itself even more obvious by darting out to inspect almost everything that moves through its area. During the heat of the day and at night, Silver-spotted Skippers roost upside down beneath leaves.

Identification	Wingspan 1¾–2⅜″. Above: brown with rectangular yellowish spots. Below: HW with large silver patch; FW with golden-orange band and spots.
Habitat	Suburbs, meadows, streamsides, and canyons.
Range	S. Canada and most of United States, excepting W. Texas and the Great Basin.
Life Cycle	Caterpillars feed on legumes (Leguminosae), especially locust (*Robinia* spp.). Adults, 1 flight to the north, 2 in the south.

Long-tailed Skipper *Urbanus proteus*

Long-tailed skippers are recognized by their relatively chunky bodies and the extended tails on the hindwings. This species also has a characteristic blue-green iridescence—a color that hints at its tropical affinities. While beautiful to the butterfly enthusiast, the Long-tailed Skipper can be a pest to the farmer. The caterpillars, called "Bean-leaf Rollers," occasionally cause significant damage to bean crops.

Identification	Wingspan 1½–2″. Above: brownish black with blue-green iridescence on body and HW; FW with row of whitish squares. Below: tan to gray. FW with white markings, HW with 2 dark bands.
Habitat	Fallow fields, gardens, and shorelines.
Range	Florida, Gulf Coast, and S. California. At times emigrating northward through Mississippi Valley and along Atlantic Coast.
Life Cycle	Caterpillars feed on legumes (Leguminosae). Adults fly all year in south.

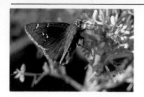

Northern Cloudywing *Thorybes pylades*

The cloudywing and duskywing skippers may justly be included with the LBJs (Little Brown Jobs) of the butterfly world. In general, however, cloudywings are characterized by relatively uniform wing surfaces marked by various glasslike flecks and bars on the forewing. Duskywings often have considerable patterning on the wing surfaces and relatively fewer glasslike flecks.

Identification   Wingspan 1¼–1¾″. Overall dark brown with glasslike flecks on FW. Below: brown with grayish wing borders. In the similar Southern Cloudywing *(T. bathyllus)*, the numerous FW markings form a loose band across the wing.

Habitat   Second growth fields, roadsides, woodland edges.

Range   Much of S. Canada and United States excepting Great Basin and central Texas, E. Colorado, Kansas, and Nebraska.

Life Cycle   Caterpillars feed on legumes (Leguminosae). Adults, 1 flight, 2 or more in South.

Juvenal's Duskywing *Erynnis juvenalis*

Identifying the various duskywings is often a job for the expert. Fortunately, several of the more common species, including Juvenal's Duskywing, are readily distinguishable if you can see them at very close range. Male Juvenal's Duskywings often alight on a conspicuous perch; from here they aggressively defend their mating grounds and seek receptive females.

Identification Wingspan 1¼–1¾". Overall: dark brown (males) to tan (females). FW with dark bands, chevrons, and glasslike scales, several in area of wing tip. Below: HW with 2 small light spots. Similar Horace's Duskywing *(E. horatius)* usually lacks the 2 light spots on the ventral HW.

Habitat Oak woodlands: edges, paths, and openings.

Range SE. Saskatchewan through central Texas; east to Atlantic Coast.

Life Cycle Caterpillars feed on oaks (*Quercus* spp.). Adults, 1 flight in spring.

Common Checkered Skipper *Pyrgus communis*

The Common Checkered Skipper flies throughout the year in its southern strongholds. Annually, during the spring and summer months, individuals from these southern populations move northward in numbers to recolonize extensive areas of the United States and southern Canada. This highly adaptive behavior, which makes them one of the more common skippers in North America, is due in large part to their ability to make use of a wide variety of host plants.

Identification
Wingspan ¾–1¼". Variable. Above: checkered black and white; males with bluish scaling on body and HW base. Below: FW as above but lighter; HW tan to whitish with black-rimmed olive bands.

Habitat
Open, brushy fields, deserts, and meadows; also disturbed areas, roadsides, and gardens.

Range
United States and S. Canada, excepting N. New England, Great Lakes region, and Northwest.

Life Cycle
Caterpillars feed on many species of mallows (Malvaceae). Adults fly all year in Deep South.

154

Common Sootywing *Pholisora catullus*

The Common Sootywing is another species that is well equipped to take advantage of areas where humans have altered the landscape. Community gardens, landfills, fallow fields, and waste places are good sites to look for the Common Sootywing. This butterfly is quick to fly and often evades capture by maneuvering close to the ground.

Identification
Wingspan ⅞–1¼". Small. Black to dark brown with tiny white spots on FW, 1 group concentrated near wing tip.

Habitat
Disturbed fields, agricultural areas, and waste places.

Range
Throughout North America from S. Canada south, but absent from Florida.

Life Cycle
Caterpillars feed on pigweeds (*Amaranthus* spp.) and goosefoots (*Chenopodium* spp.). Adults, 2 flights in the North, 3 in the South.

Least Skipperling *Ancyloxypha numitor*

This small skipper spends much of its time in the cover of tall grasses. Like most butterflies, it also visits a variety of blossoms in search of nectar. The Least Skipperling has been observed nectaring at the flowers of at least a dozen different plants, including clovers (*Trifolium* spp.), wood-sorrels (*Oxalis* spp.), and vetches (*Vicia* spp.).

Identification Wingspan ¾″. Small. Above: FW black with some orange; HW orange with dark border. Below: plain orange on both wings. Distinguished by size, deltoid wing shape when at rest, and 2-tone upper wing surfaces.

Habitat Tall grass areas, including wet meadows and marshes, field and pond edges, and hillsides.

Range E. Colorado and Dakotas south through central Texas and east to Atlantic Coast; chiefly in eastern part of its range.

Life Cycle Caterpillars feed on grasses (Poaceae). Adults, 2–4 flights.

158

European Skipper *Thymelicus lineola*

The European Skipper, a relative newcomer to North America, was introduced in Ontario in 1910. Since that time this species has extended its range considerably. This rapid expansion may be attributed in part to its reliance on crops such as Timothy. When ranchers and farmers transport bales of hay for feeding stock, European Skipper eggs may be going along for the ride. A second introduction of this butterfly was recorded in British Columbia in 1960.

Identification	Wingspan ¾–1″. Above: orange, males with tiny black stigma, females with darkened veins. Below: FW orange, HW dull mustard.
Habitat	Fields, meadows, and other grasslands.
Range	British Columbia east to Newfoundland, south in Appalachians to Carolinas.
Life Cycle	Caterpillars feed on grasses (Poaceae), especially Timothy *(Phleum arvense)*. Adults, 1 flight.

Fiery Skipper *Hylephila phyleus*

Monocultures—extensive areas where only one plant species is cultivated—create situations where another species may become a pest. Suburban lawns are prime examples of monocultures, and the Fiery Skipper can become a problem in these areas, especially in the southern states. The caterpillars forage on grasses, including common lawn species such as blue grasses (*Poa* spp.) and fescues (*Festuca* spp.).

Identification   Wingspan 1–1¼″. Short antennae. Above: yellowish orange to dark brown; males with black stigma and jagged black wing border. Below: dull orange with black spots.

Habitat   Suburban lawns, fields, and grassy edges.

Range   Carolinas west through Gulf Coast states to California. Emigrates north to S. New England, Great Lakes region, and the Southwest.

Life Cycle   Caterpillars feed on grasses (Poaceae). Adults, 3–5 flights in the South.

Yellowpatch Skipper *Polites peckius*

More than a few butterfly enthusiasts ignore skippers, lumping them together as small, unidentifiable look-alikes. While there are indeed some difficult butterflies to identify among this group, most skippers have readily recognizable field marks. The common Yellowpatch Skipper, for example, has a characteristic ventral hindwing pattern. Sometimes called *P. coras*.

Identification	Wingspan ¾–1″. Small. Above: brown with yellow markings on both wings. Below: HW with several large rectangular yellow markings which contrast with darker background.
Habitat	Fields, lawns, meadows, roadsides, and freshwater wetlands.
Range	Alberta east to Atlantic Coast and south through New England, mid-Atlantic states, and Appalachians to N. Georgia; also the northern Rocky Mountains.
Life Cycle	Caterpillars feed on grasses (Poaceae). Adults, usually 2 flights, visit a variety of blossoms, especially red clover *(Trifolium pratense)*.

Tawny-edged Skipper *Polites themistocles*

Most of our skippers are recognizable by their relatively small size, chunky bodies, and hooked antennal tips. The Tawny-edged Skipper and its relatives are members of a group called Branded Skippers, which are usually orange and black. Males often have a patch of specialized scent scales, or a stigma, on the dorsal forewing. The Tawny-edged Skipper is typical in having all these characteristics.

Identification Wingspan ¾–1″. Above: brown with orange on costal edge; male stigma is somewhat J-shaped, female with additional yellow spots. Below: FW similar to upper surface; HW mustard.

Habitat Many grassy areas, often damp; also yards and gardens.

Range Most of S. Canada and United States; absent from Far West and much of Texas.

Life Cycle Caterpillars feed on grasses (Poaceae). Adults, 2 flights.

Woodland Skipper *Ochlodes sylvanoides*

The Woodland Skipper is common in much of the West from the Rocky Mountains to the Pacific Coast. Despite its common and scientific names, this skipper is not restricted to the forests; a highly adaptable species, it also frequents scrubby hillsides, city parks, and vacant lots. It is even known to haunt the edges of tidal marshes.

Identification   Wingspan ¾–1⅛". Overall orange and black. Above: dark jagged border; males with large black stigma connecting with dark patch next to border, females with similar pattern. Below: HW often with contrasting yellow spots. Subspecies *O. s. santacruza* occurs on Channel Islands; overall darker with ventral HW chocolate with contrasting yellow spots.

Habitat   Mountain glades, hillsides, waste places, urban parks, tidal areas.

Range   Rocky Mountains to Pacific Coast and Santa Cruz Island.

Life Cycle   Caterpillars feed on grasses (Poaceae). Adults, 1 flight.

168

Northern Golden Skipper *Poanes hobomok*

The Northern Golden Skipper, also known as the Hobomok Skipper, is common throughout much of its range. An unusual aspect of this species is that there are two female forms, one similar to the male, the other a strikingly different dark form called Pocahontas. These dark females are recognizable by the purple cast to their wings, especially noticeable on the ventral wing borders. A few white spots on the forewing are also typical. Pocahontas is common in the East but rare in the Midwest.

Identification	Wingspan 1–1⅜″. Overall: orange and black. Below: large contrasting orange patch on HW and purplish cast on wing borders. Dark morph female still shows dark purple cast on wings.
Habitat	Woodland paths and openings as well as fields and meadows.
Range	E. Alberta to Maritime Provinces and south in places to Colorado and Georgia.
Life Cycle	Caterpillars feed on grasses (Poaceae). Adults, 1 flight.

Broad-winged Skipper *Poanes viator*

This large, brightly marked skipper is common in many sites along the Atlantic and Gulf coastal plains as well as in the Great Lakes region. The fortunes of the coastal populations may be tied to the aggressive phragmites reed *(Phragmites australis)*. In portions of the eastern United States, invasive phragmites colonies provide an ideal habitat for the Broad-winged Skipper.

Identification   Wingspan 1¼–1¾″. Large. Above: male dark with large contrasting orange patches; female similar with additional white patches. Below: HW reddish tan with lighter rectangular marks.

Habitat   Coastal and inland wetlands, especially with host plant grasses, sedges, and reeds.

Range   Minnesota to Maine, south to Nebraska, Texas, and Alabama, in disjunct local populations.

Life Cycle   Caterpillars feed on phragmites, wild rice *(Zizania aquatica)*, and lake sedge *(Carex lacustris)*. Adults, 1 to several flights.

Dun Skipper *Euphyes vestris*

Like many butterflies, the Dun Skipper is most easily observed while it is nectaring. This species' physical features are unremarkable, but it is often abundant around its favored blossoms. Look for Dun Skippers on blooming milkweeds (*Asclepias* spp.), dogbanes (*Apocynum* spp.), and vetches (*Vicia* spp.); they seek moisture at mud puddles and dung. Formerly known as *Euphyes ruricola*.

Identification | Wingspan 1–1¼". Plain. Overall: dark brown; mostly unmarked; females with tiny spots on FW and faint crescent on ventral HW; males with golden-orange scales on thorax and head.

Habitat | Fields and meadows, wetland edges, mountainsides, and disturbed areas.

Range | S. Canada to Florida, west to Rockies; northwestern states and Pacific Coast from Oregon to S. California.

Life Cycle | Caterpillars feed on sedges (Cyperaceae). Adults, 1 flight.

174

Roadside Skipper *Amblyscirtes vialis*

Known for their confusing similarities, twenty or so members of the genus *Amblyscirtes* occur in North America; they form a group called the Roadside Skippers. Many are relatively local in their distribution; this species, however, is widespread in North America and relatively simple to identify. This small, somewhat drab butterfly can be easily overlooked, but a diligent search within its range will often reward the butterfly seeker with his or her first view of the Roadside Skipper.

Identification	Wingspan ⅞–1″. Overall: dark. Above: with small group of tiny white spots near FW tip. Below: with violet-gray hue on HW border and FW tip.
Habitat	Woodland openings, paths, and cuts.
Range	Across Canada south to central Florida, E. Texas, Colorado, and Northwest; central California mountains.
Life Cycle	Caterpillars feed on blue grasses (*Poa* spp.) and wild oats (*Avena* spp.). Adults, 1 flight in the North, rarely 2 in the South.

Parts of Butterflies

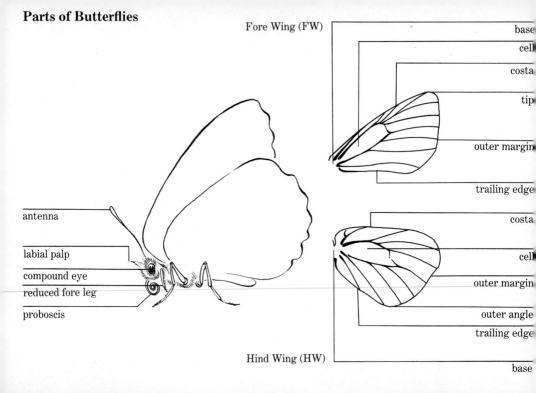

Fore Wing (FW)

base

cell

costa

tip

outer margin

trailing edge

antenna

labial palp

compound eye

reduced fore leg

proboscis

costa

cell

outer margin

outer angle

trailing edge

Hind Wing (HW)

base

fore wing

hind wing

antennae

thorax

abdomen

fore leg

tarsus

middle leg

hind leg

Butterfly Families

Every species of butterfly belongs to a family—a group of related kinds that often share recognizable traits. Learning the general characteristics of families and other groupings can be an important aid to identification and will provide you with a framework for understanding relationships between the various species. The following families are all represented in this guide:

Swallowtails (Papilionidae)

Medium-sized to large butterflies, normally with prominent tails on HW. Sexual dimorphism is common. Caterpillars have forklike "horns" (osmateria). Family includes the largest butterflies in North America and the largest species in the world, Queen Alexandra's Birdwing *(Ornithoptera alexandrae)*. The family also includes the Parnassians, a group of small alpine butterflies.

Whites and Sulfurs (Pieridae)

Small to medium-sized butterflies, overall white, yellow, or orange. Sexual and seasonal dimorphism are common. Alba (white or pale yellow) female form occurs in all species. Aggregations of males are found at mud puddles. The two subfamilies are the whites and the sulfurs.

Gossamer Wings (Lycaenidae)	Small butterflies, many with reflective scales producing bright colors such as copper, blue, green, and purple. Caterpillars of several species associate with ants. The main subfamily includes blues, coppers, hairstreaks, elfins, and the harvester. Another subfamily represented here is the metalmarks.
Snouts (Libytheidae)	Medium-sized butterflies, with one species in North America, only a dozen or so worldwide. Characterized by elongated "snout," or labial palpi. Adults mimic dead leaves.
Brushfoots (Nymphalidae)	Small to large butterflies. One of the largest families; members characterized by reduced forelegs. Includes 4 subfamilies considered here: brushfoots, leafwings, satyrs and wood nymphs, and milkweed butterflies. Brushfoots are small to large butterflies and include fritillaries (some with metallic-silver markings), crescents (smallish species with pale crescents on HW beneath), checkerspots (often with checkerspot pattern above and/or below), anglewings (with angular wing borders), tortoiseshells, thistle butterflies, admirals, and sisters. The Hackberry Butterfly represents the leafwing species. Satyrs and wood nymphs are medium-sized, brownish species of grass and woodland areas.

Milkweeds are largish butterflies of which three species occur in North America, the more common species being the Monarch and the Queen.

Skippers (Hesperiidae) A large family of small to medium-sized butterflies. Most have chunky bodies and hooked antennal clubs. Two subfamilies are represented here: the pyrgine skippers are normally dark, and some have iridescent scales or long tails; they often bask with their wings entirely open. Branded skippers are normally orange and black; many of the males have a stigma on the FW above. Branded skippers often bask with forewings and hindwings at different angles or with wings closed.

Resources

For more information about butterflies, contact these organizations.

Lepidopterists' Society
257 Common Street
Dedham, MA 02026

SASI (Sonoran Arthropod Studies, Inc.)
P.O. Box 5624
Tucson, AZ 85703

Xerces Society
10 Southwest Ash Street
Portland, OR 97204

Y.E.S. (Young Entomologists' Society)
1915 Peggy Place
Lansing, MI 48910

Glossary

Antenna
One of a pair of long, slender sensory appendages attached to the head (plural, antennae).

Batesian mimicry
An evolved likeness of an innocuous species to a different species that is unpalatable or toxic to predators.

Cell
A relatively large central area of the wing enclosed by veins.

Chrysalis
In butterflies, the pupa.

Costa
The upper or leading edge of the wing.

Diapause
A resting state in which bodily functions are minimized and energy is conserved.

Dimorphism
The occurrence of two recognizable forms in a single species.

Dorsal surface
The upper surface of a wing.

Emigration
Movement (often in large numbers) away from residential range; usually periodic but sometimes unpredictable.

Estivation
Dormancy or diapause during summer or hot weather.

Eyespot
A pattern of scales on a butterfly's wing resembling an eye.

Flight
In butterflies, a brood or generation.

Hibernation
Dormancy or diapause during winter; also called overwintering.

Host plant
The specific food of a caterpillar.

Instar
One of several stages of caterpillar growth.

Larva
Caterpillar (plural, larvae).

184

Margin
The edge of the wing.

Metamorphosis
Changes in form occurring during the life cycle; butterflies change from larva (caterpillar) to pupa (chrysalis) to adult.

Migration
A regular two-way movement to and from wintering or breeding ranges.

Osmeterium
A forklike protuberance ("horn") behind the head of swallowtail caterpillars (plural, osmeteria).

Overwinter
Hibernate.

Palpi
The two appendages on each side of the proboscis, near the mouthparts of a butterfly.

Pheromone
Sex-attractant scent molecules produced in the scales of certain butterflies.

Polymorphism
The occurrence of several distinct forms within a species; may be sexual or seasonal.

Proboscis
The elongated, extensible tube near the mouthparts of a butterfly; used to take in nectar, sap, and minerals; coiled when not in use.

Pupa
In butterflies, the stage of the life cycle occurring between caterpillar and adult; the chrysalis.

Scent scales
Specialized scales for producing and dispersing pheromones; also known as androconia.

Stigma
An area of scent scales found on the forewings of some male skippers and hairstreaks (plural, stigmata).

Submarginal
Just within the outer wing margin.

Ventral surface
The underside of a wing.

Index

190

Photographers
Gregory R. Ballmer (18, 41, 104, 105)
Doug W. Danforth (39, 44, 108, 150)
Harry N. Darrow (25, 28, 32, 37, 55, 56, 57, 61, 64, 72, 77, 85, 87, 90, 91, 100, 102, 115, 119, 121, 122, 124, 145, 151, 155, 157, 171, 173)
Thomas W. Davies (40, 45, 67, 74, 75, 94, 95, 98, 99, 103, 106, 107, 117, 118, 119, 126, 162, 169)
E. R. Degginger (79, 81)
Harry Ellis (43, 71, 141)
George O. Krizek (20, 21, 62)
C. Allan Morgan (19, 42)
Paul Opler (26, 50, 52, 58, 60, 80, 82, 83, 88, 89, 96, 127, 128, 131, 138, 146, 147, 148, 152, 153, 154, 156, 159, 161, 165, 167, 168, 174, 175, 176, 177)

National Audubon Society Collection/Photo Researchers, Inc. Danny Brass (38, 116), Ken Brate (163), Roy Coleman (24, 69, 113, 130), R. J. Erwin (34, 164), Gilbert Grant (129, 140),

Tom & Pat Leeson (31), Jeff Lepore (76), Michael Lustbader (111), Charles W. Marin (143), D. Mohrhardt (110), Kate Nicholson (144), Richard Parker (49), Stephen P. Parker (158), Rod Planck (73), J. H. Robinson (123), Kjell B. Sandved (47), John Serrao (101), Alvin E. Staffan (92), Norm Thomas (112), Larry West (22, 59), Virginia Wineland (29)

Philip Nordin (30)
Kjell B. Sandved (66)
John Shaw (27, 36, 46, 48, 54, 70, 78, 84, 86, 93, 97, 109, 120, 125, 132, 136, 137, 139)
Larry West (23, 33, 35, 51, 53, 63, 65, 68, 133, 134, 135, 142, 149, 160, 166, 170, 172)

Drawings by Mary Jane Spring

Cover photograph: Western Tiger Swallowtail by Pat and Tom Leeson/Photo Researchers, Inc.

Title page: American Painted Lady by Roy Coleman/Photo Researchers, Inc.

Spread (16–17): Clouded Sulfer by Stephen P. Parker/Photo Researchers, Inc.

All editorial inquiries should be addressed to:
Chanticleer Press
665 Broadway, Suite #1001
New York, NY 10012

To purchase this book, or other National Audubon Society illustrated nature books, please contact:
Alfred A. Knopf, Inc.
201 East 50th Street
New York, NY 10022
(800) 733-3000

This book was created by Chanticleer Press.

Founding Publisher: Paul Steiner
Publisher: Andrew Stewart

Staff for this book:

Senior Editor: Ann Whitman
Editor: Carol M. Healy
Project Editor: Ann ffolliott
Editorial Assistant: Kate Jacobs
Production: Kathy Rosenbloom, Karen Slutsky
Project Design: Paul Zakris
Photo Library: Tim Allan
Natural Science Consultant: John Farrand, Jr.

Original series design by Massimo Vignelli